NATIONAL GEOGRAPHIC

School Publishing

T0131086

THE BEAT GOES ON

PATHFINDER EDITION

By Nancy Finton

CONTENTS

2   The Beat Goes On

8   Getting Around

10   All Pumped Up

12   Concept Check

THE BEAT

Your heart beats 100,000 times a day. Each beat gives your body the oxygen it needs to survive. But what happens when the heart's not up to the job?

By Nancy Finton

Breathe. Within seconds, the oxygen you took in is cruising throughout your body.

Oxygen is carried by fast-flowing rivers of blood. What powers that bloodstream? Feel the answer for yourself. Just put your hand on your chest. Beneath the skin and bone lies a thumping, pumping muscle—your heart.

GOES ON

Mighty Muscle

Hearts work hard. Your heart beats about 90 times a minute. Each minute, in fact, this powerhouse pumps your body's entire supply of blood. That's roughly 100,000 beats a day and 2.5 billion over a lifetime. Each beat sends oxygen and disease-fighting cells throughout your body.

How does it happen? Next time you're in a pool, squeeze your hands together. Watch the water shoot up. The heart does basically the same thing. Each time it contracts, or squeezes, blood squirts through tubes called **blood vessels**.

Blood vessels come in three main types. **Arteries** carry blood away from the heart. They branch off into smaller and smaller tubes. The smallest are called **capillaries**. Capillaries have thin walls. So oxygen and other materials can travel from the blood into the cells where they're needed. **Veins** bring "used" blood back to the heart.

Broken Hearts

Most kids have healthy hearts. Trouble usually comes years later, when parts of this pumping "machine" wear out or get clogged with fat.

But there are hearts that need help sooner. Some babies are born with heart problems. Sometimes there's a hole that blood leaks through. At other times, parts of the heart are incorrectly formed—or missing. Like lifesaving mechanics, doctors step in to tune up, repair, and even replace failing hearts.

Take Heart. Ten years after his heart transplant, Brian Whitlow plays goalie on his college lacrosse team. He is a health major at Chico State University in California. He plans to work with kids in a teaching hospital.

Something's Missing!

Healthy hearts have four chambers, or sections. The two upper chambers, called **atria**, receive blood from veins. The **ventricles**, or lower sections, pump blood into arteries.

That's the plan, anyway. But Brian Whitlow's life began differently. "I was born with only one ventricle," he says. Brian's heart couldn't pump blood to his lungs to get oxygen.

But Whitlow's mother didn't give up. Nor did his doctors. When Brian was just a few weeks old, surgeons operated on him.

"We can't rebuild chambers that haven't grown," says Dr. Daniel Bernstein, one of Whitlow's childhood doctors. Instead, the surgeons rearranged his blood vessels to bypass the missing chamber.

After the operation, Whitlow's blood ran straight to his lungs. There it got oxygen, then flowed back to his heart. Whitlow's single ventricle then sent oxygen-rich blood gushing throughout his body.

That worked pretty well, Whitlow says. "I did all the normal things, including playing Little League baseball." Then things changed.

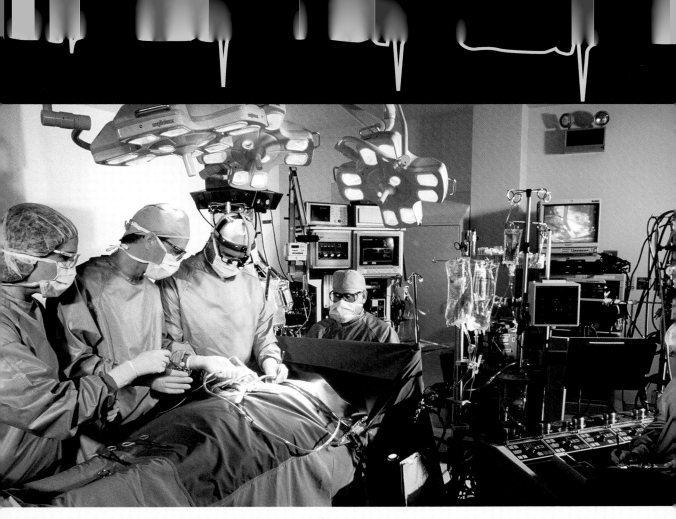

Lifesavers. New devices, procedures, and medicines have greatly reduced the risks of heart surgery.

THE GIFT OF LIFE

By the time Whitlow turned 14, his patched-up heart was pretty tired. It was too weak to keep his body strong. "I had trouble running and had to nap all the time," he remembers.

Doctors recommended a heart **transplant**. That's when a damaged body part gets replaced. "New" hearts come from people who agreed to donate them after death. "It's a difficult thing when someone has to die in order for you to live," says Whitlow. "But giving someone the gift of life is just plain wonderful."

That gift took a while to arrive. Each year about 35,000 people in this country need heart transplants. There are only a small number of replacements available. So Whitlow waited a month, then another, then another.

THE WAITING GAME

Jessica Melore knows just how Brian Whitlow felt. She had a heart attack at 16. Her left ventricle was destroyed. Melore needed a new heart. But would she get one in time?

Luckily, scientists have invented machines that can help patients like Melore lead normal lives during the long wait. Surgeons attached a doughnut-size device to Melore's heart. Powered by batteries, it did the pumping her left ventricle would have done.

The artificial pump allowed Melore to finish high school and get into college. Four days before graduation, she got word that doctors had found a new heart for her. Melore missed graduating with her class but says, "The heart was a good graduation gift."

Second Chance. After her heart transplant, Jessica Melore is an active student at Princeton University in New Jersey. She's making the most of her second chance at life.

TELLING THEIR STORIES

Today Brian Whitlow and Jessica Melore are healthy college students. They have received a lot of attention. They've appeared on talk shows and given interviews. Newspapers and magazines have written about them.

Both students hope their extraordinary stories will inspire others. They'd like to see more people arrange to have their organs donated to others after death. "We really need more organ donors," Melore says. "It's a little-known issue."

Whitlow and Melore also want to help other kids who face obstacles. "Focus on the positive things in your life," Jessica tells young audiences. "You can't change what's in the past, so make the most of the future."

THEY'VE GOT HEART

Back to Brian Whitlow. Altogether he waited 13 months for a new heart. Finally doctors called with good news. There was a heart available for him. Because hearts can't live long outside the body, Whitlow hurried to the hospital.

In the operating room, doctors hooked Whitlow to a machine that adds oxygen to blood while the heart isn't working. Then surgeons removed his damaged heart and put the new heart in its place.

Whitlow recovered quickly. "I had my surgery and started practice for high school basketball the following October," he says. But the recovery wasn't easy.

"I moved to a new high school after the transplant," Whitlow recalls. "I was taking medicine and gained 70 pounds. Kids made fun of me, and that was hard. I had to keep telling myself they didn't know my story."

Today Whitlow is on his college lacrosse team. "My teammates tease me that I'm slow, because it takes me a little longer to warm up," he says. But now he knows that they're only teasing. "Last year, they voted me the Most Inspirational Player. That felt good."

WORDWISE

artery: tube carrying blood away from the heart

atrium: upper heart chamber (plural: atria)

blood vessel: tube that carries blood

capillary: smallest blood vessel in the body

transplant: replacement of a person's damaged body part with a healthy part from another person

vein: tube carrying blood to the heart

ventricle: lower heart chamber

A Helping Heart

Each year, thousands of people wait for heart transplants. But only about 2,000 donor hearts become available. So, for many people, machines are a life saver. One machine helps a person's own heart pump blood. Another (below) is meant to replace a human heart.

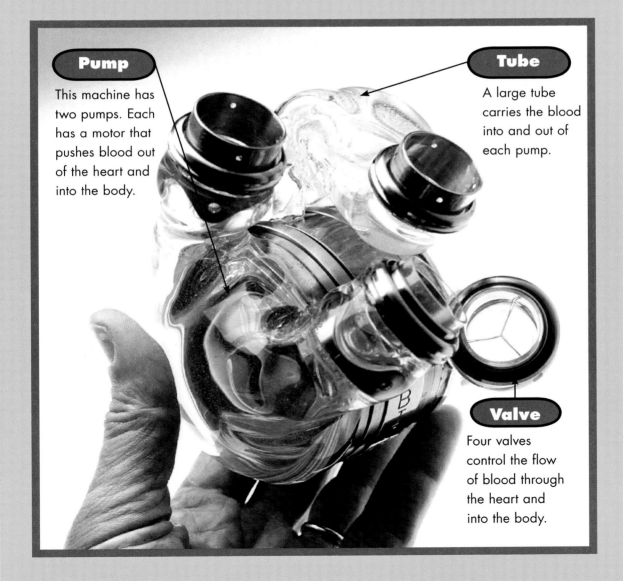

Pump

This machine has two pumps. Each has a motor that pushes blood out of the heart and into the body.

Tube

A large tube carries the blood into and out of each pump.

Valve

Four valves control the flow of blood through the heart and into the body.

Your Circulatory System:
GETTING AROUND

Your body needs oxygen to survive. Getting oxygen to all parts of your body is the job of the circulatory system. It includes your heart, several quarts of blood, and more than 96,561 kilometers (60,000 miles) of blood vessels.

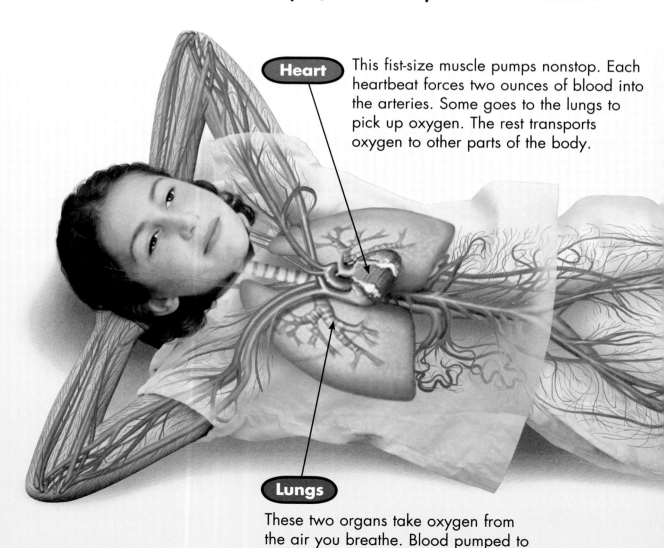

Heart

This fist-size muscle pumps nonstop. Each heartbeat forces two ounces of blood into the arteries. Some goes to the lungs to pick up oxygen. The rest transports oxygen to other parts of the body.

Lungs

These two organs take oxygen from the air you breathe. Blood pumped to the lungs then picks up the oxygen.

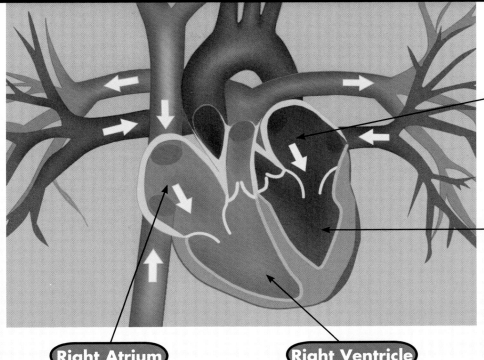

Left Atrium

Receives oxygen–rich blood from the lungs

Left Ventricle

Pumps oxygen–rich blood throughout the body

Right Atrium

Gets "used" blood from the body

Right Ventricle

Pumps blood to the lungs to pick up oxygen

Artery

Blood flows away from the heart through blood vessels called arteries. They generally carry oxygen–rich blood, which is bright red.

Vein

Blood flows back to the heart through blood vessels called veins. "Used" blood has a dark color.

ALL PUMPED UP

Each beat of your heart pumps blood through your body. That blood is packed with oxygen. When you exercise, your body needs more of this oxygen-rich blood to keep going. So when you're active, your heart really gets pumping.

How does exercise affect your heartbeat? Find out by doing this easy experiment. You may want to work with a partner.

Predict

1 Will exercise make your heart beat more or less?

Test

2 Place two fingers on your wrist. Find a spot where you feel a sort of thumping. That's your pulse. Each thump represents a heartbeat.

3 Count how many thumps you feel in 30 seconds. Write the number down.

4 Now run in place as fast as you can for two minutes.

5 As soon as you finish, place two fingers on your wrist and count your pulse again. How many times does it thump in 30 seconds now?

Conclude

6 Did your heart beat more or less after you exercised?

7 Why did your heartbeat change when you exercised?

THE HUMAN HEART

It is time to find out what you have learned about the human heart.

1 How does your heart move blood through your body?.

2 What is the difference between arteries and veins?

3 What do the atria and ventricles do?

4 How does a temporary artificial heart help people survive?

5 How might stories about heart transplants inspire other people?

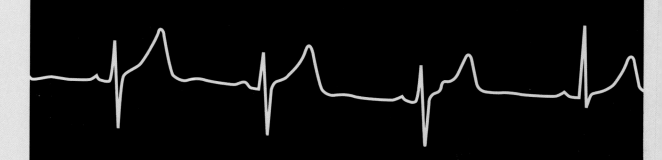